THE PASTOR'S GUIDE TO A.I. 2.0

Leading Your Church in the Age of Intelligent Technology

Kevin Crawley
Pastor | Author | Kingdom Entrepreneur

❀ Created with Vellum

THE PASTOR'S GUIDE TO A.I. 2.0

LEADING YOUR CHURCH IN THE AGE OF INTELLIGENT TECHNOLOGY

KEVIN L CRAWLEY

ABSKO CONSULTING

To the pastors on the front lines — you didn't sign up to be technologists, but you were always called to be leaders. This book is for those brave enough to embrace what's next without losing what matters most.

CONTENTS

A NOTE BEFORE WE BEGIN

When I wrote the first Pastor's Guide to AI, I remember exactly where I was sitting. I was in my office in Phoenix at about 6 in the morning, coffee in hand, trying to find the words to explain to pastors that something was coming. Something big. Something that the church could not afford to ignore. I was writing about something on the horizon.

That was not very long ago. And yet, looking back, it feels like a lifetime.

The AI landscape has shifted so dramatically, so quickly, that much of what was cutting-edge two years ago is now standard. What was speculation is now reality. AND what the church was debating whether to engage with, most pastors are now using — whether they realize it or not.

This book is not a revision of the first one. It is a new field guide for a new moment. The church is no longer behind on AI. In fact, a recent survey found that 91% of church leaders now support the use of AI in ministry, and 61% use it weekly or daily. That is remarkable. **[1]** But here is the part that keeps me up at night: 73% of those same churches have no AI policy whatsoever. **[1]**

That is not an AI problem. That is a leadership problem. AND that is exactly why I wrote this book.

We are going to go deeper this time. We are going to address the real concerns — the ones pastors actually bring up when no one is recording them. We are going to get practical in ways the first book only scratched the surface of. And we are going to take a first look at what is coming next: agentic AI, the technology that does not just answer your questions but actually completes entire tasks on your behalf.

I have been a pastor long enough to know that the church has never been well-served by leaders who bury their heads in the sand when the world changes around them. AND I have been a Kingdom Entrepreneur long enough to know that every shift brings opportunity — if you are willing to lean in.

My prayer is that by the time you close this book, you will feel equipped, not overwhelmed. Inspired, not anxious. And above all, I pray that you will see what I have come to believe more firmly than ever: that the greatest days of the local church are still ahead, and the tools to help you get there are already in your hands.

Let's get to work.

Apostle Kevin L Crawley

Phoenix, Arizona

INTRODUCTION: IT'S NOT JUST HERE — IT'S RUNNING

I want you to do something with me for a moment. Think back to three years ago. Think about what your phone could do, what your computer could do, and what technology in general looked like in your world. Now I want you to think about today — about the tools you have access to right now that you did not have then. The difference is staggering, is it not?

Now here is a more sobering thought: three years from today, the gap between now and then will be even wider.

We are living in the most rapid period of technological change in human history. I do not say that for dramatic effect. I say it because understanding the pace of change is essential for every leader who wants to navigate it wisely. AI is not a trend. It is not a fad. It is not something that is going to peak and then fade the way social media platforms come and go. Artificial Intelligence is an infrastructure shift — the kind that changes everything around it permanently.

Think of electricity. When it was first introduced, people debated its usefulness. Some thought it was a novelty. Others were afraid of it. A few visionaries understood that it was not just a new product but a new foundation — one that would eventually power everything. AI is that kind of shift. It is not just a new app or a new tool. It is a new layer of capability that is being built into every platform, every device, and every industry on earth.

Including the church.

WHERE WE ARE RIGHT NOW

Here are some numbers that should stop you in your tracks. A nationwide survey of church leaders conducted in 2025 found that two-thirds of Protestant pastors now use AI to help prepare their sermons. **[1]** Not dabble. Not occasionally. Regularly. The study also found that 61% use AI tools weekly or daily in their ministry, **[1]** and that number is growing at a pace that surprises even the researchers tracking it.

That means the question is no longer "should the church use AI?" The church already is. The question now is: are we using it wisely, intentionally, and in a way that honors God and serves our people well?

My answer to that question is: we are getting there. But we are not there yet.

THE GAP THAT MUST BE CLOSED

The most alarming statistic I came across while researching this book is not about AI adoption rates. It is about the gap between adoption and preparation. While 91% of church leaders support AI use in ministry, 73% have no AI policy for their staff. **[1]** None. That means thousands of well-meaning pastors and ministry leaders are using powerful technology without any framework for how it should be used, what its limits are, or how to protect their congregation from the risks that come with it.

That gap is not a technology problem. It is a leadership problem. And the church has always been at its best when leaders step into the gaps rather than waiting for someone else to fill them.

This book is your guide to closing that gap.

WHAT THIS BOOK WILL DO

We are going to cover a lot of ground together. We will start by addressing the real concerns — the ones that keep pastors up at night, the theological questions, the fears about authenticity, and the legitimate worries about misinformation. I want to meet those concerns head-on, not dismiss them.

Then we are going to build a foundation. You will learn how to create an AI policy for your church, how to choose the right tools for your specific ministry context, and how to use AI in a way that extends your reach without sacrificing your soul.

From there, we will go deep on the practical applications: preaching, content creation, pastoral care, administration, discipleship, and more. By the time we are done, you will have a toolkit that any church can implement, regardless of size, staff, or technical background.

And finally, we are going to look at what is coming next. Specifically, we are going to talk about agentic AI — the next frontier that most pastors have not even heard of yet, but which is already beginning to transform how organizations operate.

My goal is simple: I want you to finish this book feeling like a leader, not a follower. In a moment when the world is racing ahead, the church cannot afford to be the last one to show up.

Let's begin.

PART ONE
CLEARING THE AIR

Addressing the Concerns That Keep Pastors Up at Night

CHAPTER 1
LET'S ADDRESS THE ELEPHANT IN THE ROOM

I have been in a lot of conversations about AI with pastors and church leaders over the past few years. And I have noticed something interesting. The conversations almost always start the same way. Someone brings up AI, and before the discussion even gets going, someone else says, "But what about..." And then one of a handful of concerns gets raised.

Well, I want to honor those concerns. They are not bad concerns. They are not ignorant concerns. Many of them are deeply thoughtful and spiritually grounded. But here is the thing — they often end up functioning as conversation-stoppers rather than conversation-starters. And that is a problem, because the conversation needs to happen.

So let us have it. Right here, right now. Let us put the biggest concerns on the table and talk about them honestly.

CONCERN #1: WILL AI REPLACE THE HOLY SPIRIT IN MINISTRY?

This is the concern I hear most often, and I want to honor it, because it comes from the right place. Pastors are people who have built their lives on the conviction that ministry is not just a human enterprise. It is a Spirit-led endeavor. And there is a very real fear that leaning on AI tools could, over time, crowd out that Spirit-led dimension.

Let me be direct about this: that fear is legitimate. But it is not a reason to avoid AI. It is a reason to be intentional about how you use it.

Here is the truth. AI cannot pray. AI cannot fast. AI cannot hear from God. AI cannot sense the shift in a room when the Spirit of God is moving. AI cannot look a grieving widow in the eyes and carry the weight of her loss. AI cannot sit with a prodigal son and speak life into him in the way that only someone who has been redeemed can do. These things are not features that AI is missing because it has not been updated yet. They are characteristics of the human soul, set apart by God for sacred work.

What AI can do is help you research, write, organize, and communicate. And here is the key distinction: a hammer does not replace a carpenter. It extends one. AI is a tool. The anointing is still yours. AND it always will be.

"AI can serve as a tool that enhances our ministry, amplifies our skills, and extends our reach. It can never, and should never, overshadow or replace the unique gifts, insights, and anointing that God has placed within you."

CONCERN #2: WHAT ABOUT THEOLOGICAL ACCURACY?

This one is important, and I want to give it the weight it deserves. AI language models are trained on enormous amounts of text from across the internet, books, articles, and other sources. That means they have absorbed a wide range of theological perspectives — some orthodox, some not. When you ask an AI tool a theological question, it does not have a doctrinal statement it answers from. It generates a response based on patterns in its training data.

The risk here is real. There have already been documented cases of AI-generated sermon content that included fabricated quotes, misattributed scriptures, or theologically imprecise statements that sounded convincing but were not accurate. One widely reported case involved an AI-generated sermon that fabricated a quote attributed to the medieval philosopher Maimonides — convincing enough that it went unnoticed by many listeners.

But here is the thing: this is not a reason to reject AI. It is a reason to never stop being the theologian in the room.

Think about your Bible app. It can pull up any translation, any commentary, any concordance result you want in seconds. But if you pull up the wrong translation for the wrong context, or if you take a commentary's opinion as settled doctrine, the problem is not the app. The problem is the lack of discernment in how you used it. AI works the same way. It is a powerful resource that requires a skilled, theologically grounded leader to evaluate and apply it wisely.

Always verify. Always check scripture in context. Always run AI-generated theological content through the filter of your own study, prayer, and doctrinal understanding. That discipline has not changed. What has changed is the speed of the resource.

CONCERN #3: IS AI-GENERATED CONTENT AUTHENTIC?

I hear this one a lot, particularly around sermon preparation. The question usually goes something like this: "If I use AI to help me write my sermon, is it really my sermon? Am I being authentic to my congregation?"

This is a good question. And I want to flip it slightly.

Let me ask you: when you read a commentary to help shape your understanding of a text, is your sermon less authentic? When you listen to another pastor's message for inspiration, does that undermine your integrity? When your editor cleans up the manuscript of your book, does the book stop being yours?

Every communicator uses resources. The question is not whether you use them. It is whether what you ultimately deliver is genuinely yours — shaped by your study, your prayer, your relationship with the people you serve, and your encounter with God's word.

AI should be part of your process, not a replacement for it. Use it to generate an outline and then rework it in your voice. Use it to suggest illustrations and then choose the ones that resonate with your experience. Use it to do research and then bring your own theological weight to the conclusions. The sermon that comes out the other side should still be undeniably yours. If it is not, the problem is not AI. It is the absence of your own investment.

CONCERN #4: WHAT ABOUT PRIVACY AND DATA SECURITY?

This concern has grown significantly as AI tools have become more sophisticated, and it deserves practical attention. When you type content into an AI tool, you need to understand where that data goes. Some tools store your conversations. Some use your inputs to train future models. And some have data sharing policies that are less than transparent.

When it comes to your congregation, this becomes especially serious. If someone shares a prayer request with you and you paste it into an AI tool to help craft a response, that person's private information may now be stored on a server somewhere. That is not just a tech concern. It is a pastoral and ethical one.

The solution is not to avoid AI. It is to use AI responsibly. We will cover this in detail when we talk about creating a church AI policy. But for now, the principle is simple: never put sensitive personal information about your congregation into AI tools that are not specifically designed with privacy protections in mind.

THE BOTTOM LINE

Every single one of these concerns is valid. I want you to know that I have not dismissed them or glossed over them. But here is what I have found after years of studying this intersection of faith and technology: the concerns, when held properly, do not lead us away from AI. They lead us toward wisdom in how we use it.

The church has always had to navigate new tools with discernment. This moment is no different. The calling is not to have all the answers before you start. The calling is to lead. NOW is the time to do exactly that.

CHAPTER 2

WHAT THE CRITICS GET RIGHT (AND WHERE THEY STOP SHORT)

I want to be fair to the critics, because some of them are raising important points. And honestly, I think the church needs to hear those points before it can move forward wisely.

Let me tell you about a church in Finland that decided to do something bold. They held an entire Sunday service created entirely by AI. The sermon was generated by AI. The music was AI-composed. The visuals were AI-generated. Even the on-screen representations of Bible characters were AI avatars. The service ran. People showed up. And attendance actually went up for that week.

But here is what happened in the aftermath. Attendees reported that while the service was technically impressive, something felt missing. They described it as lacking warmth. Several people said it felt hollow. The experience confirmed something that I believe most of us know intuitively: you can replicate the structure of worship without capturing the spirit of it.

That is what the critics get right. Efficiency without anointing is just efficiency. And in the church, efficiency without anointing is not ministry. It is production.

THE RISK OF OUTSOURCING YOUR PREPARATION

Here is another thing critics are right about. When you outsource your sermon preparation entirely to AI, you are not just saving time. You are skipping the process. And the process matters.

There is something that happens when a pastor wrestles with a text, when you sit with a passage for days, when you pray over it and argue with it and let it confront you before you stand up to preach it. That process does not just produce a better sermon. It produces a better pastor It deepens your own walk with God in ways that no shortcut can replicate.

I believe that one of the real spiritual dangers of AI is not that it will corrupt the church from the outside. It is that it could quietly atrophy the spiritual muscles of leaders who stop doing the work of seeking God because they can get a sermon outline in thirty seconds.

That is a legitimate concern. AND every pastor who picks up an AI tool should carry that concern with them — not as a reason to put the tool down, but as a discipline for how to pick it up.

WHERE THE CRITICS STOP SHORT

And yet — and this is where I depart from the most cautious voices — the answer to that concern is not to put the tools down. The answer is to use them differently.

Think about every major technological advance the church has navigated. Radio. Television. The printing press. The internet. In each case, there were voices in the church warning that the new technology would compromise the integrity of ministry. And in each case, they were not entirely wrong. Every tool can be misused. Every tool carries risk.

But consider what the church would have lost if it had

listened to those warnings and stepped back from the microphone, the printing press, and the internet. Think about the souls that were reached through radio ministries in the early 20th century. Think about the millions who have been saved, discipled, and encouraged through Christian television. Think about the global reach of online ministry that became a lifeline during a worldwide pandemic.

The church that engaged those tools did not lose its soul. It extended its reach. AND the same can be true with AI.

The critics are right that AI can be used badly. They are right that it carries risks. But they stop short when they imply that the only safe option is abstention. Because abstention has a cost too. The church that does not learn to use these tools wisely will increasingly find itself unable to communicate effectively in a world that is being shaped by them.

A BALANCED POSTURE

What I am advocating for is a posture of engaged discernment. Not breathless excitement and not fearful avoidance. Engaged discernment means we stay in the room. We learn the tools. We evaluate them theologically and practically. We establish guardrails. We make mistakes and course-correct. And we keep our eyes on the mission.

That is what wise leaders do. That is what Kingdom-minded leaders do. And that is what this book is designed to help you do.

CHAPTER 3
THEOLOGY OF TECHNOLOGY — DOES GOD CARE ABOUT YOUR TOOLS?

I want to take a step back from the practical and spend a chapter on the theological. Because I believe that before a pastor can confidently and faithfully engage any new technology, they need a biblical framework for why doing so is not just acceptable but actually part of faithful stewardship.

So let me ask the question directly: Does God care about your tools?

I believe — with everything in me — that the answer is yes. And I believe the Bible gives us a clear foundation for that answer.

GOD IS NOT AFRAID OF CAPABILITY

One of the most fascinating passages in Scripture, when you think about it through the lens of technology, is found in Exodus 31. God is giving Moses instructions for the building of the tabernacle, and He says something remarkable. He tells Moses that He has filled a man named Bezalel with the Spirit of God, with wisdom, understanding, and knowledge, in all kinds of craftsmanship. God then commissions Bezalel to design, to create, to build.

What strikes me about this passage is that God did not just give Bezalel a vision. He gave him skill. Capability. Craftsmanship. And then He filled that capability with His Spirit and put it to work for Kingdom purposes.

God has never been afraid of human capability. In fact, He is the one who gave it to us. The image of God imprinted on us — the Imago Dei — includes creativity, problem-solving, and innovation. When human beings create tools and technologies that make them more effective, that is not something separate from God's design. In many ways, it is an expression of it.

PAUL WOULD HAVE USED EVERYTHING AVAILABLE TO HIM

Think about the Apostle Paul. Paul was arguably the greatest evangelist and church planter in the history of Christianity. He was also, by every account, incredibly strategic about using whatever tools and advantages were available to him.

Paul leveraged his Roman citizenship as a legal tool. He used his knowledge of Greek philosophy to connect with Athenian intellectuals in Acts 17. He understood how to adapt his communication style for different audiences. When he could not be present in person, he wrote letters — a technology of his day — that traveled across the Roman Empire and connected churches spread thousands of miles apart.

I have said this before and I will say it again: I believe Paul would have used every AI tool available to him if he were alive today. Not to replace his prayer life. Not to substitute for his encounter with the risen Christ. But to extend his reach, accelerate his communication, and remove as many barriers between himself and the people he was trying to reach as possible.

The only thing Paul and the early apostles had in common with us today is the guidance and leading of the Holy Spirit.

The tools available to us now far exceed anything they had access to. AND I believe they would not waste that advantage.

STEWARDSHIP IS A TECHNOLOGY ISSUE

Here is a frame I want to offer you: stewardship. We talk about stewardship of finances, stewardship of time, stewardship of our gifts and calling. But stewardship also applies to the tools God has allowed to exist in our moment in history.

You did not choose to be a pastor in the age of AI. But you are. And that means you have been entrusted with a set of resources and capabilities that pastors of every previous generation did not have. The question of stewardship is: what are you going to do with them?

I do not think God is honored when His leaders ignore available resources out of fear or tradition. I do not think He is glorified when the church falls behind the culture in its ability to communicate the timeless Gospel. I think He is glorified when Spirit-led leaders take what is available to them and consecrate it for His purposes.

That is what we are called to do with AI.

THE ONE THING AI WILL NEVER REPLACE

Let me be clear about this, because it is the theological anchor of everything else in this book.

AI can generate words. It cannot generate conviction. AI can research theology. It cannot know God. AI can craft a sermon. It cannot carry the weight of one. AI can produce content. It cannot carry anointing.

These are not features AI is missing because it is still developing. They are the irreducible, sacred dimensions of a life surrendered to God and called into ministry. Your voice carries something that no language model can replicate: the testimony of your journey, the weight of your calling, and the authority that comes from time spent in the presence of God.

That is not going anywhere. AI can help you do more. It can help you reach further. But the "you" in that sentence — the pastor, the shepherd, the Spirit-led leader — that is still entirely yours. AND it is still the most irreplaceable thing in your ministry.

"Technology changes. The mission does not. And the one who carries the mission — the anointed, called, Spirit-led leader — is still the most powerful force in any community."

PART TWO
BUILDING YOUR FOUNDATION
ESTABLISHING WISE FRAMEWORKS BEFORE YOU BUILD

CHAPTER 4
BEFORE YOU BUILD, READ THE BLUEPRINT — CREATING YOUR CHURCH AI POLICY

Let me give you a number that should make every church leader uncomfortable: 73%.

That is the percentage of churches that currently use AI with no policy in place to govern how it is used. [1] No guidelines for staff. No guardrails around sensitive data. No framework for what AI can and cannot do in their ministry context. Seventy-three percent.

I understand how this happens. Leaders are busy. AI tools are easy to sign up for and start using immediately. The benefits are obvious and fast. The risks are less visible, at least at first. And creating a policy sounds like something that requires a board meeting, a committee, and a six-month process nobody has time for.

But here is the reality: you do not need a perfect, comprehensive policy on day one. You need a starting point. And I am going to give you one right now.

WHY A POLICY MATTERS

A church AI policy is not bureaucracy for its own sake. It is pastoral care for your congregation and protection for your ministry. Consider what can go wrong without one.

A well-meaning staff member copies a prayer request into an AI chatbot to help craft a pastoral response. That prayer request — which may include details about someone's marriage crisis, health struggle, or financial collapse — is now stored on a commercial server. The person who shared it had no idea.

A volunteer uses an AI tool to draft a children's ministry curriculum. The AI includes a theological statement that contradicts your church's doctrinal position. The curriculum goes through without review and gets taught to a hundred kids.

A pastor uses an AI tool to generate a sermon illustration. The AI fabricates a quote attributed to a historical figure. The congregation hears it. Someone Googles it. And suddenly the pastor's credibility is on the line.

These are not hypothetical scenarios. Versions of all of these have already happened in churches across the country. A clear, simple policy prevents most of them.

THE FIVE PILLARS OF A CHURCH AI POLICY

Pillar 1: Transparency with Your Congregation

Your people deserve to know when and how AI is being used in your ministry. This does not mean a legal disclaimer at the bottom of every email. It means a culture of openness. If your bulletin is designed with the help of AI, that is fine — and you do not need to announce it every Sunday (that would get old fast). But if your sermon was substantially drafted by AI, your congregation should know that.

Transparency is not just about managing expectations. It is

about trust. And trust is the currency of pastoral ministry. Be open about your tools.

Pillar 2: Protecting Member Data

This is non-negotiable. Create a clear rule: no personally identifiable information about your congregation members goes into any AI tool. No names paired with prayer requests. No financial giving information. No counseling session details. No health or family crisis information.

If you are using AI to help craft communications to specific individuals, remove identifying details before you input the information. Use placeholders. Then personalize the final output yourself before it goes out. This is not inconvenient. It is responsible.

Pillar 3: Theological Review of All AI-Generated Content

Every piece of AI-generated content that reaches your congregation — sermons, Bible study materials, social media posts, emails, children's curriculum — should go through a theological review before it goes out. That review does not have to be lengthy. But it needs to happen.

Designate someone on your team whose job it is to verify that AI-generated content aligns with your doctrinal standards, uses scripture accurately and in context, and represents your church's voice and values. This person does not have to be a theologian. They just need to be someone who knows your church and knows how to spot something that does not sound right.

Pillar 4: Define What Needs Human Authorship

Not everything can or should be AI-assisted. Define clearly for your staff and volunteers which functions require human authorship. For most churches, that list will include pastoral care responses to personal crises, counseling-related communications, public statements on sensitive issues, and key leadership communications like vision messages from the senior pastor.

Everything else is potentially fair game for AI assistance,

but "assistance" is the operative word. A human being should always be in the loop.

Pillar 5: Stay Current and Review Regularly

The AI landscape is changing fast. A policy written today may need significant updates in twelve months. Build in a regular review cadence — at least annually — where your leadership team evaluates your AI policy in light of new tools, new risks, and new best practices. The church that set a policy and forgot about it is almost as vulnerable as the church that never set one.

A SIMPLE STARTING POINT

If you are not sure where to begin, here is a simple one-page framework you can adapt for your church right now. The same 2025 survey that uncovered the policy gap also found that 87% of church leaders said they would be willing to invest in AI training for their ministry teams. **[1]** That means the will is there. What is often missing is a practical starting point. I have included a full template in Appendix A at the back of this book. Start there. Adapt it. Make it yours. And then share it with your team.

The goal is not a perfect policy. The goal is a starting point that protects your people, guides your staff, and creates a culture of intentionality around the tools you are already using.

CHAPTER 5
KNOW YOUR TOOLS — A PRACTICAL GUIDE TO TODAY'S AI LANDSCAPE

One of the most common things I hear from pastors when I talk about AI is some version of this: "I know I should be using it, but I am not sure where to start. There are so many tools out there. How do I know which ones are actually worth my time?"

It is a fair question. (And I know — there are approximately one million AI tools being launched every week. I am exaggerating, but only slightly.) The AI tool market has exploded, and navigating it can feel like walking into a hardware store the size of a football field when all you need is a hammer.

Let me simplify it for you. The tools that matter for ministry fall into a few clear categories. Here is what each category is, what it does, and which tools I recommend in each one as of the time this book was written. These will evolve — but the categories will not change nearly as fast as the tools within them.

CATEGORY 1: GENERAL-PURPOSE AI ASSISTANTS

These are the Swiss Army knives of AI — tools that can help you write, research, brainstorm, summarize, translate, and much more. Think of them as an incredibly smart assistant who never sleeps, never gets annoyed by your questions, and never asks for benefits.

ChatGPT (OpenAI) — Still the most widely known and one of the most capable general-purpose AI tools available. The paid version (ChatGPT Plus, powered by GPT-4) is significantly more powerful than the free version and is worth the cost for ministry use. It is excellent for sermon research, drafting communications, brainstorming series ideas, and creating outlines.

Google Gemini — Google's AI assistant, which integrates directly with Gmail, Google Docs, and other Google Workspace tools. If your church already uses Google Workspace, Gemini can be a natural fit because it can help you draft emails, summarize documents, and create content without leaving the tools you already use.

Microsoft Copilot — Microsoft's AI assistant integrates with Word, Outlook, Excel, and Teams. If your church uses Microsoft 365, Copilot can help you draft sermons in Word, summarize emails in Outlook, and analyze data in Excel. This integration can significantly streamline your workflow.

Claude (Anthropic) — Known for being particularly strong at nuanced writing, theological discussion, and handling long documents. Many pastors and writers prefer Claude for its thoughtful, balanced approach to complex topics.

CATEGORY 2: MINISTRY-SPECIFIC AI TOOLS

These are tools built specifically for church and ministry contexts. They often have theological guardrails built in and are designed around the specific workflows pastors deal with.

Sermon Shots — An AI tool that helps create short-form video clips from sermon content. Great for social media repurposing.

SermonAI — Designed specifically for sermon preparation, with features for creating outlines, illustrations, and application points.

Church.tech — A suite of AI tools built specifically for churches, covering everything from member engagement to content creation.

CATEGORY 3: WRITING AND COMMUNICATION TOOLS

Grammarly — Uses AI to help you write more clearly and professionally. Excellent for staff emails, church communications, and anything you want to make sure is grammatically clean and clearly expressed.

Jasper — AI writing tool that is particularly good for marketing-style content: website copy, social media, email campaigns, and fundraising appeals.

CATEGORY 4: IMAGE AND VISUAL CREATION

Midjourney — One of the leading AI image generation tools. You can create custom sermon series artwork, social media graphics, and visual illustrations by simply describing what you want in plain language. The quality has become remarkably realistic.

DALL-E (OpenAI) — Integrated into ChatGPT Plus, this

allows you to generate images directly within your AI conversations. Great for quick visual content creation.

Canva AI — Canva, already a popular tool for church graphics, has integrated AI features that make creating professional-looking visuals even faster and honestly even more fun.

CATEGORY 5: TRANSLATION AND ACCESSIBILITY

Wordly — Real-time AI translation for live events. Attendees can listen to your sermon or worship service in their native language through an app on their phone. For churches in diverse communities, this is a game-changer. Full stop.

DeepL — One of the most accurate AI translation tools available. Great for translating written content into multiple languages.

A SIMPLE RULE FOR CHOOSING TOOLS

With so many options, here is a simple rule I use: start with one. Pick one general-purpose AI tool — I usually recommend ChatGPT or Claude to start — and use it consistently for thirty days. Learn its strengths. Notice where it falls short. Then add a second tool for a specific need once you know what that need is.

The biggest mistake leaders make is trying to learn five tools at once. That leads to overwhelm, inconsistency, and usually abandonment. One tool, used well and regularly, will transform your workflow. Start there.

PART THREE
AI AT WORK IN YOUR MINISTRY

Practical Applications Across Every Area of Church Life

CHAPTER 6
THE SERMON, REIMAGINED

I want to start this chapter with a confession.

When I first started using AI for sermon preparation, I was cautious about it in a way that I now recognize was more pride than wisdom. Part of me felt like using AI was somehow admitting that I needed help. That it meant my own study and preparation were not enough. I had to get honest with myself about where that feeling was coming from.

And what I discovered — when I was honest enough to look at it — was that the resistance was not theological. It was ego. I did not want to feel like I needed a tool. I wanted to feel like the sermon flowed directly from heaven through me with no assistance required. Which sounds very spiritual until you realize you have been reading three commentaries a week for the past fifteen years.

The truth is, we have always used resources. Commentaries. Bible dictionaries. Other preachers' sermons for inspiration. Research databases. Illustrations from books and podcasts. None of that made our preaching less authentic. It made it better. AI is the most powerful resource in that category I have ever come across. And once I got over myself and started using it properly, it changed my preparation process in ways that genuinely surprised me.

WHAT AI CAN DO FOR YOUR SERMON PREPARATION

Let me walk you through what a well-integrated AI-assisted sermon preparation process can look like. I want to be clear: this is not a process where AI writes your sermon. It is a process where AI functions as your research assistant, your brainstorming partner, and your fact-checker, while you remain the preacher.

Step 1: Use AI as a Research Partner

Begin with the text. After your own initial study and prayer, bring your passage to an AI tool and use it to deepen your research. Here are some prompts that have been genuinely useful in my own preparation:

- "What are the key theological themes in [passage]? What do major commentaries say about this text?"
- "What is the historical and cultural context of [passage]? What would the original audience have understood?"
- "What are some common misinterpretations of this passage that I should be aware of and address?"
- "What are some cross-references in scripture that illuminate this passage?"

The AI can pull together research from a wide range of sources and present it to you in a fraction of the time it would take to compile manually. But remember: evaluate everything. Read the actual passages it references. Check the theological claims. Use this as a starting point, not a final answer.

Step 2: Use AI as a Brainstorming Partner for Illustrations

Illustrations are often the hardest part of sermon preparation. Finding a story or example that genuinely illuminates the truth you are trying to communicate, without being clichéd or forced, is an art. AI can help.

Try prompts like:

- "I am preaching about [topic/text]. What are some real-world contemporary examples or stories that illustrate [key truth]?"
- "Give me ten different ways to illustrate the concept of [theological concept] for a general audience."
- "What are some historical examples from church history that illustrate [theme]?"

The AI will generate options. You will choose the ones that resonate with your own experience, your congregation's context, and your personal voice. The illustration that lands is almost always the one that the pastor connects with personally — and that is still your job.

Step 3: Use AI as a Counterargument Challenger

This is one of my favorite uses of AI in sermon prep, and most pastors have not thought of it. Before you finalize your message, present your main argument to an AI tool and ask it to push back.

Try something like: "Here is the main point I am making in my sermon this Sunday: [your main point]. What are the most common objections someone might have to this? What might a skeptic or a new believer struggle with?"

This exercise does two things. First, it helps you anticipate questions and objections from your congregation so you can address them proactively. Second, it strengthens your own thinking by forcing you to engage with the strongest counterarguments before you get up to preach.

Step 4: Keep Your Voice in the Center

This bears repeating. Whatever AI helps you research,

brainstorm, or outline, the sermon that comes out the other end needs to sound like you. The personal stories are yours. The convictions are yours. The pastoral heartbeat for the people sitting in front of you is yours.

Read your AI-generated outline out loud before you refine it. Anything that does not sound like something you would say needs to be rewritten in your voice. This step is not optional. It is the step that transforms an AI-assisted sermon into your sermon.

A WORD ABOUT FULL AI-GENERATED SERMONS

I want to address this directly, because some pastors are doing it. Using AI to generate an entire sermon, from text selection to application, without meaningful personal input or theological review, is not something I recommend. The concerns I raised in earlier chapters apply here: the risk of theological error, the loss of personal authenticity, and the spiritual atrophy that comes from skipping the process.

That said, I understand that there are emergency situations. You are sick. A family crisis hits on Saturday night. Your guest speaker cancels last minute. In those cases, an AI-drafted sermon with rapid theological review and personalization is infinitely better than nothing. Use your judgment. But let those be exceptions, not habits.

CHAPTER 7
CONTENT THAT MULTIPLIES — SOCIAL MEDIA, EMAIL, AND BEYOND

One of the most exhausting challenges of modern church leadership is content. The expectations placed on churches in the digital age are enormous. You are expected to have a consistent social media presence across multiple platforms, a regular email newsletter, a podcast or YouTube channel, a website that is current and compelling, and increasingly, short-form video content.

For most churches, especially smaller ones, this is simply not sustainable without AI. You do not have a content team. You have a pastor and maybe a part-time communications person (who is also the worship leader, the event coordinator, and the guy who fixes the sound system on Sunday mornings). And that combination cannot produce the volume and variety of content the digital world demands.

AI changes that equation entirely. AND here is where it gets genuinely exciting for churches of every size.

THE CONTENT ECOSYSTEM STRATEGY

Here is the framework I teach to churches of every size. It is built around a simple idea: one sermon, many pieces of content.

Every Sunday, you preach a message. That message is the content hub. Everything else flows from it. AI is the multiplier that turns one piece of source material into a week's worth of communication.

Here is how it works in practice:

Step 1: Capture Your Sermon Content

If you preach from a manuscript or detailed notes, you already have the source material. If you preach more conversationally, record your sermon and use an AI transcription tool (like Otter.ai or Descript) to turn it into text. You now have raw material.

Step 2: Feed It to Your AI Tool

Paste your sermon transcript or notes into your AI tool and use prompts like:

- "Based on this sermon, write five Instagram captions of 150 characters or less that highlight the key ideas."
- "Write a Facebook post (300-400 words) that summarizes the main point and invites people to watch the full message."
- "Write a 250-word email newsletter introduction that teases the main point of this message and encourages people to join us this week."
- "Identify five quotable one-liners from this sermon that would work well as text graphics."
- "Write a 60-second script for a short-form video (Reels / TikTok / YouTube Shorts) based on the strongest point of this message."

Step 3: Review, Personalize, and Publish

Everything the AI produces is a draft. Your communications person — or you, if that is still you — reviews each piece, adjusts the voice to match your church's tone, and makes sure it is accurate and on-brand. Then it goes out.

What might have taken a full day of creative work now takes a couple of hours. And the quality is often better, because you are building from the strongest material you have — your preaching — rather than trying to generate content from scratch.

EMAIL THAT GETS OPENED

AI is also transforming how churches communicate by email. Not just the newsletter, but the individual touchpoints that make people feel seen and known.

Consider these use cases:

- First-time visitor follow-up emails that feel personal, not templated
- Birthday and anniversary messages to congregation members
- Donor appreciation letters that go beyond the standard tax receipt
- Event invitations that tell a story rather than just listing logistics
- Mid-week encouragement emails tied to Sunday's message

For each of these, AI can produce a strong first draft in seconds. The key is to personalize and make sure the final product sounds like a human wrote it — because a human reviewed and refined it.

A NOTE ON AUTHENTICITY IN DIGITAL COMMUNICATION

Your congregation can tell the difference between communication that is real and communication that feels produced. The goal is not to eliminate that human quality. The goal is to have AI handle the volume so your human quality can shine through more consistently.

Think of it this way: without AI, you might send a generic monthly email to your whole congregation. With AI, you can send a personalized, content-rich weekly touchpoint that actually says something. The AI made the volume possible. Your pastoral heart still makes it meaningful. That is not a compromise — that is the right use of the tool.

CHAPTER 8
CARING FOR YOUR PEOPLE AT SCALE

Pastoral care is the beating heart of ministry. The hospital visit. The phone call at midnight. The conversation after service that turns into a three-hour counseling session. The follow-up text to someone you noticed was not themselves on Sunday. These moments are irreplaceable. They are the difference between a church and a religious organization.

They are also not scalable. Not in the traditional sense.

If you are a pastor of 200 people, you can probably maintain meaningful personal connection with most of your congregation. If you are a pastor of 500 people, it becomes a stretch. If you are leading a church of 1,000 or more, individual pastoral care at the level your people need and deserve becomes mathematically impossible without a large, well-coordinated team.

AI does not solve this problem by replacing pastoral care. It solves it by helping you not lose people in the cracks. And losing people in the cracks is something every pastor I know carries as one of their deepest burdens.

AI FOR CONNECTION AND FOLLOW-UP

Here is one of the most powerful applications of AI in the pastoral care context: using it to make sure no one falls through.

Imagine a system where every first-time visitor to your church receives a thoughtful, personalized follow-up email within 24 hours. Not a generic "thanks for coming" message, but a message that references the service they attended, reflects the message you preached, and invites them back with warmth and specificity.

Imagine a system where every prayer request submitted through your app or website generates a draft pastoral response that a staff member reviews and sends within hours, rather than days.

Imagine being able to send a personal check-in message to every member you have not seen in three or four weeks — prompted by your church management system and drafted by AI, personalized by a staff member, and sent with a human signature.

None of this replaces the pastoral visit. But it ensures that the pastoral visit gets to the right people at the right time, because the lower-touch touchpoints have been handled well.

AI-POWERED DISCIPLESHIP TOOLS

Several ministries are now developing AI-powered discipleship tools — tools that can answer common faith questions, provide scripture for specific situations, suggest resources for spiritual growth, and guide new believers through the basics of the Christian faith, available 24 hours a day.

Think about the person who comes to faith at 11 PM on a Tuesday and has burning questions that cannot wait until Sunday. Or the new believer in a time zone on the other side of the world who connected with your ministry online and needs guidance. An AI discipleship tool can provide immediate, theologically grounded responses while connecting them with a human pastor for deeper follow-up.

This is not a replacement for community. But it is a bridge. AND bridges save lives.

THE GUARDRAILS THAT MUST BE IN PLACE

I want to be direct here, because this area carries real risk. AI should never be used to provide pastoral counsel on serious mental health crises, suicidal ideation, abuse situations, or major life decisions without immediate human escalation. Full stop.

Any AI-assisted pastoral care system needs a clear protocol for when a situation escalates to human intervention, and that protocol needs to work quickly. The cost of getting that wrong is too high.

Additionally, as discussed in the chapter on church AI policy, no personal information shared in a pastoral care context should be stored in AI systems without clear disclosure and consent. Your congregation needs to trust you with the most vulnerable parts of their lives. Protect that trust fiercely.

CHAPTER 9
RUNNING YOUR CHURCH LIKE A LEADER, NOT A MANAGER

There is an old saying that leaders work on the system while managers work in it. Most pastors spend far too much time working in the system — answering emails, scheduling meetings, managing logistics, tracking data, writing reports. These things matter. But they are not your highest contribution. Your highest contribution is vision, leadership, and pastoral presence.

Now here is the good news: AI has the potential to give you back enormous amounts of time and mental energy by handling the administrative weight that eats up your week. Let me show you how.

EMAILS AND COMMUNICATIONS

The average pastor spends multiple hours per week on email alone. AI can dramatically reduce that time by helping you draft responses faster, summarize long email threads, and maintain consistent communication quality even when you are tired or pressed for time.

Try this: instead of staring at a blank reply, give your AI tool the context of the email and what you want to say, and let it draft a response. You edit it, personalize it, and send it. What might take fifteen minutes of labored typing now takes three minutes. Do that ten times a day, and you have reclaimed over two hours.

MEETING PREPARATION AND FOLLOW-UP

AI transcription and summarization tools are transforming how leadership teams handle meetings. Here is a practical workflow:

1. Record your elder board meeting or staff meeting.
2. Run the recording through an AI transcription tool.
3. Feed the transcript to your AI assistant and ask it to generate a summary, action items, and follow-up tasks.
4. Review, edit, and distribute the summary within hours of the meeting.

This alone can cut hours from your administrative week. Meetings become more accountable because the action items are captured. Follow-up becomes consistent because the reminders are automated. And the record of decisions is clear and accessible.

BUDGET AND FINANCIAL ANALYSIS

AI tools, particularly those integrated with spreadsheet software like Microsoft Excel or Google Sheets, can help you make sense of financial data quickly. Giving trends. Budget variances. Year-over-year comparisons. Instead of spending an hour preparing a financial report for your elder board, you can ask your AI tool to analyze the data and present the key takeaways.

This does not replace your finance team or your accountant. But it helps you walk into financial conversations as an informed leader rather than someone who is still trying to understand what the numbers mean. That is a significant difference.

GRANT WRITING AND FUNDRAISING

If your church operates any kind of community program or non-profit ministry, grant writing is probably one of the most time-consuming and daunting tasks you face. AI is remarkably good at helping draft grant proposals, because grant writing follows a predictable structure that AI handles well.

Give your AI tool the details of your program, your community impact, your budget needs, and the specific grant you are applying for, and ask it to draft a proposal. The first draft will need significant editing, but it gives you a powerful starting point rather than a blank page. And a blank page is the enemy of the busy pastor.

VOLUNTEER MANAGEMENT AND SCHEDULING

Several church management platforms now have AI features built in that can help optimize volunteer scheduling, send automated reminders, track participation patterns, and identify volunteers who may be at risk of burnout or disengagement. If your church management system does not have these features yet, it will soon.

CHAPTER 10
TRAINING, TEACHING, AND DISCIPLESHIP AT A NEW LEVEL

One of the most exciting — and honestly most underutilized — applications of AI in ministry is in the area of teaching and discipleship content. The ability to create customized, high-quality educational content for any audience, at any level, in a fraction of the time it used to require, is genuinely transformative for the local church.

I get excited talking about this one, because I have seen it in action. Churches that used to take months to build out a curriculum series are now doing it in weeks. Small churches with no paid curriculum staff now have the ability to produce content that looks and reads like something from a major publishing house. That matters.

BUILDING BIBLE STUDY CURRICULUM

Creating a Bible study series from scratch is a significant undertaking. Research, writing, formatting, discussion questions, application points — for a thorough series, you might be looking at weeks of work. AI can compress that timeline dramatically.

Here is a sample prompt to get you started:

"Create a six-week Bible study series on the Book of Ruth. For each week, include: (1) the scripture reading, (2) a brief background and context section (3 paragraphs), (3) five discussion questions that range from observation to application, (4) a key theological takeaway, and (5) a practical application challenge for the week."

What you get back will be a strong first draft that your teaching pastor or small group director can refine, personalize, and contextualize for your congregation. The bones are done. The finishing work is what makes it yours.

CUSTOMIZING FOR DIFFERENT AUDIENCES

One of the most powerful things AI enables is audience customization. A sermon series on prayer can be explored at a children's level, a new believer level, a mid-level discipleship level, and an advanced theological level — and each version should feel native to its audience.

Creating four versions of the same content used to take four times the work. With AI, you create one strong base and then use prompts to adapt it. "Rewrite this content for middle school students." "Adapt this discussion guide for new believers who have no church background." The AI handles the rewrite. You handle the review.

LEADERSHIP DEVELOPMENT TRACKS

Growing the next generation of leaders in your church is one of your most important long-term investments. It is also one of the things that most often gets pushed aside by immediate operational demands. I know — I have lived that tension too.

AI can help you build structured leadership development tracks without starting from scratch. Define the competencies you want to develop in your emerging leaders. Ask AI to help you build a curriculum around those competencies. Use it to create reading lists, reflection questions, mentoring guides, and assessment tools.

This does not replace the relationship-based mentoring that is at the heart of leadership development. It creates the framework within which that mentoring can happen more intentionally. AND that is a win worth working toward.

PART FOUR
THE NEXT FRONTIER

Where AI Is Headed and How to Lead Your Church There

CHAPTER 11
WELCOME TO THE AGE OF THE AI AGENT

I want to take you somewhere most pastors have not gone yet. Not because it is too technical or too complicated, but because it is so new that most of the conversation has not made it out of the tech world yet.

This is the part of the book where I introduce you to what is likely the most significant development in AI since ChatGPT launched and changed the conversation for everyone: agentic AI.

Now — before your eyes glaze over at the word "agentic" — let me start with a story. Because that is always the best way to understand something new.

THE DIFFERENCE BETWEEN A CONSULTANT AND AN EMPLOYEE

Imagine you are a pastor who needs to put together a church anniversary event. You have two options.

Option one: you hire a consultant. You sit down with the consultant, you answer their questions, you give them information about your church, your vision for the event, and your budget. They give you their best advice and recommendations. You take notes. Then they leave. And you go implement everything yourself.

That is what AI has been until recently. You ask it questions. It gives you great answers. You go do the work.

Option two: you hire an event coordinator. You tell them what you want. They ask clarifying questions. Then you go back to leading your church, and they go handle it. They book the venue. They coordinate the vendors. They manage the timeline. They send the communications. They show up on the day of the event ready to run it. You check in at key points, approve decisions, and show up to your own anniversary service.

That is agentic AI.

Agentic AI is AI that does not just answer your questions but actually completes tasks on your behalf. It can take an objective, break it down into steps, execute those steps using a variety of tools and resources, make decisions along the way, and deliver a completed result — often with minimal human intervention in the middle.

Now tell me that does not change everything.

WHY THIS IS DIFFERENT FROM EVERYTHING BEFORE

Up until now, AI has been reactive. You prompt it, it responds. You ask a question, it answers. Every interaction requires a human to initiate it.

Agentic AI is proactive. You can give it a goal, set it in motion, and it will work toward that goal — making decisions, taking actions, using tools, and updating you on its progress.

Consider some ministry examples of what this could look like:

Agentic AI for Member Care

You configure an AI agent with access to your church

management system. It monitors attendance patterns. When it notices that a member has not attended in three weeks, it automatically drafts a personalized check-in message, sends it to your pastoral care coordinator for review and approval, and logs the interaction. You never had to notice the absence or initiate the follow-up. The agent did.

Agentic AI for Content Creation

You record your Sunday sermon. An AI agent picks up the recording, transcribes it, identifies the key quotes and themes, drafts the week's worth of social media content, schedules it across your platforms according to your preferences, creates the email newsletter, and generates a short-form video script for your communications director. By Monday morning, the content ecosystem for the week is largely done.

Agentic AI for Administrative Tasks

You configure an AI agent to monitor your church's giving dashboard. When it identifies that quarterly giving is tracking below budget, it automatically prepares a financial summary report for your elder board meeting, suggests two or three communication strategies based on your church's history, and drafts a stewardship message for your review. You are informed and equipped before you even knew there was a problem.

THE QUESTION EVERY PASTOR IS THINKING

I know what you are thinking right now. And I want to address it directly.

"Is this safe? Should I be comfortable with AI agents making decisions and taking actions on behalf of my church without me in the loop for every step?"

Here is my honest answer: it depends entirely on how you set it up. Agentic AI is not a runaway system. The most well-designed implementations of agentic AI include what are called "human-in-the-loop" checkpoints — moments where the agent pauses and waits for human review and approval before proceeding.

Think of it like a very capable executive assistant who has been with you for years and knows your preferences deeply. You trust them to handle routine matters independently. But for anything significant, they bring it to you first. That is what well-designed agentic AI should feel like.

The key is to define upfront: what can the agent do autonomously, and what requires your approval? The clearer those boundaries are, the safer and more effective your agentic AI implementation will be.

WHERE AGENTIC AI IS RIGHT NOW

As of the time this book is being written, agentic AI is still early in its development, particularly for ministry applications. Tools like OpenAI's Operator, Microsoft's Copilot Agents, and various platforms built on top of foundational AI models are making agentic capabilities available to everyday users for the first time.

You do not need to implement agentic AI in your church tomorrow. But you need to know it is coming. You need to understand what it does. And you need to be positioned to evaluate it wisely when it becomes more accessible and more ministry-specific — which will happen faster than most people expect.

PREPARING YOUR CHURCH FOR THE AGENT ERA

Here is what I recommend for forward-thinking church leaders right now:

1. Learn the concept. Understand what agentic AI is and how it differs from the AI tools you are currently using. This chapter is a starting point.
2. Identify your highest-value repetitive processes. What does your church do on a regular basis that follows a predictable pattern? Those are the best candidates for eventual agentic AI implementation.
3. Strengthen your data foundations. Agentic AI works best when it has access to clean, well-organized data. If your church management system is a mess, start cleaning it up now.
4. Build a culture of intentional AI use. Churches that are already using AI thoughtfully and with clear policies will be best positioned to integrate more powerful agentic tools when they arrive.
5. Stay curious. Read. Follow the developments. Talk to other church leaders who are engaging with these tools. The leaders who stay curious and keep learning will be the ones who lead their congregations wisely into the next season.

CHAPTER 12
YOUR CHURCH IN 2027 — THINKING AHEAD WITHOUT LOSING YOUR ANCHOR

I want to spend this final chapter thinking out loud with you about the future. Not because I have a crystal ball. Not because anyone does. But because one of the most important things a leader can do is cultivate a posture toward the future that is both hopeful and discerning.

As a Kingdom Entrepreneur, I have always believed that the future belongs to those who prepare for it — not anxiously, not naively, but with both eyes open and faith leading the way. So let me tell you what I see coming.

WHAT IS COMING

Voice AI Is About to Get Very Personal

Right now, most AI interactions happen through text. You type, it responds. But voice AI is advancing at a remarkable pace. Tools like ElevenLabs are already enabling highly realistic AI-generated voices. Real-time voice AI assistants are becoming sophisticated enough to handle complex conversations. And the combination of voice AI with agentic capabilities means that in the near future, your "AI assistant" may literally be a voice that you speak to, that speaks back, that takes notes, schedules your meetings, drafts your communications, and manages your calendar.

For ministry, this has significant implications. Imagine a pastoral care line that is staffed 24 hours a day by an AI voice assistant that can listen, pray with callers, and connect them with a human pastor when needed. Imagine a member being able to call a church number at 2 AM in crisis and hearing a warm, present voice that does not dismiss them but engages, supports, and ensures follow-up. That is not science fiction. That is next year.

AI-Generated Video Is Arriving

The ability to generate realistic, high-quality video from text prompts is advancing faster than most people realize. We are approaching a moment where a pastor will be able to generate a short devotional video in their own appearance and voice without standing in front of a camera — simply by typing a script and selecting their AI avatar.

The applications for church communications, online outreach, and discipleship content are enormous. So are the ethical questions around authenticity and disclosure. Once again, the technology is not the problem or the solution. The leadership and the policy around it are what will determine whether it serves the mission.

Real-Time Translation Breaks Language Barriers

This is one I am personally excited about. Real-time AI translation is already good enough that a multilingual church can provide simultaneous language access to every service through a simple app on attendees' phones. But the next generation of this technology will be even more seamless — integrated directly into hearing devices, smart glasses, and venue systems.

For churches in diverse communities, this technology could fundamentally transform who can access and belong to your congregation. The language barrier, one of the oldest walls dividing the global church, is genuinely falling. AND that is worth celebrating.

AI-Assisted Counseling Triage

Mental health needs in the church have never been higher. Pastoral counseling demand regularly exceeds what any church staff team can handle. AI-assisted triage tools are beginning to emerge that can help identify the urgency and nature of someone's need, match them with appropriate resources, and ensure no one waits too long without some form of response.

This is a sensitive area that requires careful implementation and clear boundaries. But the potential to extend your pastoral care capacity without compromising its quality is real.

THE ANCHOR THAT DOES NOT MOVE

I want to close this chapter — and this section of the book — with something I feel strongly about.

Every technology I have described in this chapter will change. The tools that exist today will be replaced by better tools. The capabilities that feel cutting-edge right now will feel basic in three years. And three years from now, there will be capabilities that we cannot fully anticipate today.

That kind of acceleration can feel disorienting. It can create anxiety in leaders who feel like they are always behind, always catching up, always one technology cycle away from becoming irrelevant.

But here is the anchor: the mission does not change.

We are still called to make disciples. We are still called to gather the church, preach the word, baptize believers, serve the vulnerable, and advance the Kingdom of God in every community we are planted in. That calling is not technology-dependent. It is Spirit-dependent.

The tools will change. They always have. The church that

used the printing press is the same church that used radio and television and the internet and now AI. And through every technological shift, the thing that made the church the church — the presence of the Holy Spirit, the proclamation of the Gospel, the community of believers in covenant with one another — that remained constant.

It will remain constant through this shift too.

Your job is not to keep up with every new tool. Your job is to lead your people with wisdom, courage, and discernment. The tools exist to serve that calling. Not the other way around.

CONCLUSION: THE ANOINTING IS STILL YOURS

I want to take you back to where we started.

When I wrote the first Pastor's Guide to AI, the goal was simple: help church leaders understand that AI was real, that it was coming to ministry whether they invited it or not, and that they could engage it faithfully. I wanted to lower the fear and raise the curiosity.

That book found its audience. And over the months that followed, I had more conversations with pastors about AI than I have had about almost any other topic in recent memory. Some of those conversations were energizing — leaders who were already experimenting and excited about the possibilities. Some were sobering — leaders who felt left behind and anxious about a world that was changing faster than they could track.

What strikes me now, looking back at all of those conversations, is that the most confident and effective leaders I spoke with had something in common. They were not necessarily the most technically sophisticated. They were not the ones with the biggest budgets or the most staff. They were the ones who had a clear sense of what they were called to do and were willing to use every available resource to do it better.

That is what I hope this book has given you. Not just information about AI tools, but a framework for thinking about them. Not just techniques, but a theological foundation for why engaging these tools is an act of stewardship. Not just a list of what AI can do, but a clear and confident understanding of what it cannot do and will never do — the things that are irreducibly, gloriously yours.

AI cannot pray. It cannot be filled with the Spirit. It cannot carry the weight of a calling. It cannot look into the eyes of a grieving person and carry their sorrow with them. It cannot preach with the authority that comes from time spent with God.

You can.

That is not something AI will ever take from you. It is something that no advancement in technology can replicate. The anointing is still yours. The calling is still yours. The mission is still yours.

The tools are just better now.

Use them wisely. Use them boldly. Use them in service of a Kingdom that has been advancing since before the first printing press and will keep advancing long after whatever comes after AI.

The church is not behind on this moment. You are reading this book. You are learning. You are preparing to lead.

That is exactly where you are supposed to be.

To the glory of God and the advancement of His Kingdom,

Pastor Kevin Crawley

APPENDIX A: CHURCH AI POLICY TEMPLATE

CHURCH NAME — Artificial Intelligence Use Policy
Effective Date: _______________ | *Reviewed Annually by:* _______________

Purpose

This policy establishes guidelines for the use of artificial intelligence (AI) tools by staff, volunteers, and ministry leaders of [Church Name]. Our goal is to leverage AI technology in ways that enhance our ministry, protect our congregation, and honor our values and doctrinal commitments.

Approved Uses of AI

- Sermon research, outline development, and illustration brainstorming (with pastoral review)
- Church communications: social media, email newsletters, website content
- Administrative tasks: scheduling, meeting summaries, report drafting
- Curriculum development for adult, youth, and children's ministry (with theological review)
- Translation of materials for multilingual ministry contexts

- Image and graphic creation for ministry communications

Restrictions

- No personally identifiable member information (names, prayer requests, health or financial details) shall be entered into any AI tool
- No pastoral counseling responses shall be sent without human review and personalization
- All AI-generated theological content must be reviewed for doctrinal accuracy before distribution
- AI-generated sermons or teaching may not be delivered without substantial pastoral input, personalization, and theological review

Disclosure Guidelines

Staff should be transparent about AI use in significant communications. When AI has substantially contributed to a piece of content, that use should be noted where appropriate.

Data Privacy

Staff shall review the privacy policies of any AI tool before use. Sensitive member information shall never be used as AI input. Questions about appropriate data use should be directed to [designated contact].

Review and Updates

This policy will be reviewed annually and updated as the AI landscape evolves. Questions or concerns should be directed to [designated church contact].

APPENDIX B: RECOMMENDED AI TOOLS BY MINISTRY FUNCTION

Sermon Preparation & Theology

- ChatGPT Plus (research, brainstorming)
- Claude (theological discussion, nuanced writing)
- SermonAI (sermon-specific features)
- Logos Bible Software (AI-enhanced biblical research)

Content Creation & Communications

- Jasper (marketing and social content)
- Grammarly (writing clarity and grammar)
- Canva with AI (graphic design)
- Descript (video/audio editing with AI transcription)

Social Media & Video

- Sermon Shots (sermon clip creation)
- Opus Clip (auto-generates short clips from long videos)

- Later or Buffer (AI-assisted social scheduling)

Translation & Multilingual Ministry

- Wordly (real-time in-person translation)
- DeepL (written translation)
- Google Translate (general quick translation)

Image Generation

- Midjourney (highest quality image generation)
- DALL-E via ChatGPT Plus (integrated image creation)
- Adobe Firefly (commercially safe image generation)

Administration & Operations

- Microsoft Copilot (integrated with Office 365)
- Google Gemini (integrated with Google Workspace)
- Otter.ai (meeting transcription and summarization)

APPENDIX C: SAMPLE PROMPTS FOR PASTORS

Sermon Preparation Prompts

- "I am preaching from [passage]. What are the key themes, historical context, and main interpretive positions on this text?"
- "Give me 10 contemporary illustrations for a sermon on [topic] that would resonate with a general adult audience."
- "Here is my main point: [your point]. What are the strongest objections or questions someone might have to this?"
- "Create a five-part sermon series outline on [topic] with a title, key scripture, and main point for each week."

Content Creation Prompts

- "Based on this sermon excerpt [paste excerpt], write five Instagram captions of 150 characters or less."

- "Write a 300-word Facebook post that summarizes the key takeaway from this message and invites people to attend church this Sunday."
- "Create five quotable one-liner statements from this content that would work as text-over-image graphics."
- "Write a 60-second video script based on the strongest point of this sermon."

Administrative Prompts

- "Here are the notes from our leadership meeting [paste notes]. Summarize the key decisions and list the action items with owners and deadlines."
- "Draft a thank-you letter to donors who gave over $500 this year. Our church's mission is [mission]. The tone should be warm and personal."
- "Create a volunteer job description for a [role] at a mid-sized evangelical church."

Pastoral Care Prompts

- "Draft a warm, pastoral follow-up email to a first-time visitor. They attended a service about [topic]. Keep it brief, genuine, and inviting."
- "Write a check-in message for a congregation member who has been absent for several weeks. Keep it brief, warm, and non-intrusive."

APPENDIX D: GLOSSARY OF AI TERMS (PLAIN LANGUAGE)

Artificial Intelligence (AI): Technology that enables computers to perform tasks that typically require human intelligence, such as understanding language, recognizing patterns, and generating content.

Large Language Model (LLM): The type of AI that powers tools like ChatGPT and Claude. It has been trained on enormous amounts of text and can generate human-like writing and answer questions.

Generative AI: AI that can create new content — text, images, video, audio — rather than just analyzing existing content.

Prompt: The instruction or question you give to an AI tool. The quality of your prompt significantly affects the quality of the AI's response.

Agentic AI: AI that can complete multi-step tasks autonomously, making decisions and taking actions to accomplish a goal, rather than just responding to individual questions.

Human-in-the-loop: A design approach where AI systems include checkpoints for human review and approval before taking significant actions.

AI Hallucination: When an AI tool generates information that sounds confident and plausible but is actually inaccurate or fabricated. This is why fact-checking AI outputs is essential.

Natural Language Processing (NLP): The branch of AI that enables computers to understand, interpret, and generate human language.

Real-time Translation: AI-powered translation that happens instantly, allowing people to hear or read content in their own language as it is being delivered.

AI Policy: A set of guidelines that governs how AI tools are used within an organization, including what is permitted, what is restricted, and how data is protected.

ENDNOTES

[1] Exponential AI NEXT, AIforChurchLeaders.com, and ChurchTechToday.com. "The 2025 State of AI in the Church Survey Report." December 2025. Available at churchtechtoday.com/2025-state-of-ai-in-the-church/. The survey of church leaders across the United States reported that 91% of church leaders support the use of AI in ministry; 61% use AI tools weekly or daily (up from 43% in 2024); 73% of churches have no AI policy in place for staff; 64% of pastors regularly use AI for sermon preparation; and 87% expressed willingness to invest in AI training for their ministry teams.

[2] Yle News (Finnish Broadcasting Company). "Finnish Lutheran Church Holds AI-Generated Sunday Service." March 4, 2025. yle.fi. Also reported by Christian Today and TechXplore, March 2025. The service was held at St. Paul's Lutheran Church in Helsinki, Finland, and was attended and covered by multiple international news outlets.

[3] Multiple news outlets have documented incidents of AI tools generating sermon content and theological material containing fabricated quotations attributed to historical figures, including medieval scholars and theologians. These documented cases — some of which went undetected by initial reviewers — underscore the critical necessity of theological review and fact-checking for all AI-generated ministry content. Pastors are strongly advised to verify all historical attributions and quotations before use in any public ministry context.

ABOUT THE AUTHOR

Kevin L. Crawley is a pastor, Ministry Futurist, and entrepreneur who believes the Church should lead culture — not lag behind it. He serves as a senior apostolic leader of Empowerment Hub Phoenix alongside his wife, and has spent years helping pastors and ministry leaders navigate the intersection of faith, technology, and leadership.

He is the author of *Stunted Growth, The Pastor's Guide to AI,* and *The Pastor's Guide to AI 2.0,* and hosts **The WarRoom** — a podcast built to equip modern pastors and leaders with the strategies, insights, and tools they need to thrive.

Kevin is passionate about making AI accessible, practical, and Kingdom-centered for the local church.

Connect with Kevin:

Website: KevinCrawley.com

Instagram: @kevincrawley

Join the **Skool Community** at:
https://www.skool.com/ai-for-the-called-and-curious-1269/about

Podcast: The WarRoom — available on all major platforms

www.ingramcontent.com/pod-product-compliance
Lightning Source LLC
LaVergne TN
LVHW022312200926
848481LV00004B/1308
* 9 7 9 8 9 8 8 6 4 3 9 5 1 *